Fertil

M000311840

Charles Siegchrist

CONTENTS

Introduction

Readily available plant and animal residues can supply an extremely valuable source of plant foods that over the years will reward you with better soil, better crops, and reduced grocery and fertilizer bills. Some, such as those leaves on your maple trees or the grounds from your morning coffee, come free for the taking. But before you can expect great results, you must put some time and effort into improving the soil. This bulletin is aimed at helping you do just that at as low a cost as possible.

Building a garden soil with organic amendments is an investment in the future. The first year soil-building is undertaken, you probably won't see dramatic results, but as the initial applications of compost, mulch, green manures, and soil amendments break down, you'll see changes. The soil will become darker in color, easier to work, and faster to warm in the spring. You'll see more earthworms, and the structure should improve to a crumbly condition. Your gardens will need fewer waterings. Soil tests should start to return with advice to use less fertilizer and less lime, saving you money.

Feed the soil well and it will reward you many times over in the years ahead.

The Makeup of Soil

Most good soils contain consistent percentages of air, water, organic matter, and soil organisms.

Soil organisms —

Organic matter 10%

Water 25%

Rock particles 40%

Air 25%

Building Blocks

Soil is composed of rock particles, soil organisms, water, air, and organic matter in various stages of decomposition. In a fertile agricultural soil, the ratios of these parts might well appear as in the diagram on the previous page.

Rock Particles

In terms of both weight and volume, rock particles are the major constituent in soil. The action of glaciers, water, and wind reduces rocks in size by weathering them into tiny fragments. The size of these fragments, also known as soil grains, can vary greatly. The coarsest are known as gravel, followed in order by sand, silt, and clay.

Soils are characterized by the proportions of each of these particle sizes they contain. Sandy soil, with its large, coarse particles, won't retain water very long. For this reason it warms up quickly in spring and can generally be worked earliest. It also quickly shows the effects of water shortage during periods of dry weather. Unfortunately, as water drains out of sandy soil, it often takes essential plant nutrients from applied fertilizers along with it.

Soils containing a large proportion of silt or clay, with their minute rock particles, have more surface area for water to cling to and thus are better able to remain moist. This is advantageous during a dry spell but can mean a later start for the spring gardener, and clay soils may hold on to water too well. Also, clay soils in

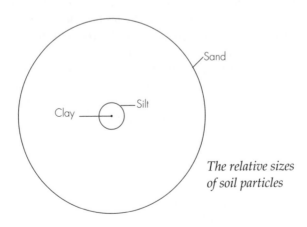

The relative sizes
of soil particles

particular are prone to compaction if worked or walked on when wet; compaction makes soil harder to work and squeezes air out of the soil, so plant roots and soil organisms can't breathe.

Soils that contain a balanced mix of these particle types are called loam. They are easiest to work and have a good balance of other properties, such as retaining a good amount of rainfall while draining quickly enough so they don't stay soggy.

Air

Under ideal conditions, about 25 percent of a soil's volume will be air. This circulates through passages between the soil crumbs, aiding in the decomposition of organic matter and sustaining soil life. Earthworms rise to the surface after a rainstorm because the soil channels and pores that supply air for them to breathe have become waterlogged. Roots, too, need air for a good exchange of oxygen and carbon dioxide, so most crops grow poorly on soggy, air-deprived soil. Many beneficial microbes are dependent on an abundant air supply as well; they are killed off when soil remains soggy for long periods.

Water

Another 25 percent of good soil is water. Its value is obvious to anyone who's ever witnessed an extended drought. Though preventing plants from wilting is the most obvious benefit, this soil component plays other essential roles. Water facilitates the liberation of plant food from rock particles and decaying organic matter. Water also transports nutrients to the roots of growing plants and disperses the nutrients through the soil.

Water occupies part of the pore space between the soil crumbs, mostly as a thin film surrounding individual particles. It also becomes absorbed into organic matter, which is capable of holding several times its own weight in water.

Organic Matter

The decomposing remains of plant and animal matter within the soil are organic matter. When this material has decomposed to a fine, dark substance in which the source of origin can no longer be determined, it's termed humus. All humus is derived from organic matter, but not all organic matter has decomposed to the point that it can be considered humus.

Though organic matter is only a small fraction of most soils, it's far more important than you might imagine. It helps hold rock particles loosely together into crumbs, or aggregates, making soil easier to work. It aids in the retention of water and in the passage of air. Further, it provides food for organisms living within the soil.

Soil Organisms

Numerically, the most common constituent of fertile soil is the population of organisms living within it. But their generally microscopic size makes them the smallest portion of the soil by volume.

One of the largest and most beneficial life-forms in the soil is the common earthworm. Worms consume large quantities of organic matter and soil particles, modifying them through digestion into nutrient forms readily available to growing plants.

Microorganisms also help liberate plant foods from organic matter and soil particles. Many different types and species of microorganisms live in the soil; these include bacteria, fungi, actinomycetes, and algae. The smallest, bacteria, are usually the most numerous; billions live in just a square inch of healthy soil.

Not all soil life wears a white hat, however. Moles ruin lawns, white grubs gnaw off roots, and harmful species of nematodes burrow into roots to strangle the flow of nutrients and water. Other microbes are responsible for wilt diseases and root maladies. Fortunately, soils also contain beneficial nematodes that feed on the harmful species (you can even purchase these if your soil is deficient), plus microorganisms that feed on disease-causing organisms.

Manipulating the Components

The character of a soil can be altered by manipulating any one or all five of its constituents. The most long-lasting way to do this is by adding organic matter.

While you can't change the amounts of sand and clay, you can make a soil with too much of either behave more like ideal loam by working in quantities of organic matter. Well-decomposed organic matter will improve the structure of either light sandy soils or heavy clay soils. In coarse sand, decaying organic matter will act as a sponge for water, reducing this soil's tendency to dry quickly. A clay soil will be lightened and made less soggy, as organic matter helps open up air channels and maintain a more porous structure.

Cultivation temporarily increases the amount of air in soil. But cultivating too frequently can destroy the beneficial crumbly structure you work so hard to create and maintain. To get the maximum aeration effects with the minimum amount of cultivation, keep the soil well stocked with humus.

Excessive water can be removed by drainage; in small areas this is easiest to achieve by elevating the soil surface a few inches with raised beds. The improved soil structure created by adding organic matter will create more tiny channels for water to drain through. Too little rainfall can be made up by irrigation, and improving the water-absorbing and water-holding ability of light sandy soils by adding organic matter will enable you to make more efficient use of irrigation water.

Organic matter is the key constituent to managing the soil for fertility. It supplies small amounts of all major nutrients and a balanced dose of all essential micronutrients or trace elements. It feeds the soil organisms that convert nutrients into the forms plants can absorb. It acts as a gentle magnet to keep nutrients from washing away and to keep them within easy reach of plant roots. Properly used, it can benefit all other aspects of soil as well: its drainage and aeration, its capacity to hold water, its microbial life and earthworm population, and its workability.

If soil life is deficient, you can inoculate the soil with fresh compost or aged (not dehydrated) manure. While you can purchase microbial treatments or mail-order earthworms, soil conditions must be good for them to survive. Usually increasing the amount of organic matter in soil will automatically increase the populations of earthworms and beneficial microbes.

The Soil in High Gear

Under natural conditions, the organic content of soil is at least maintained and frequently increased. The carpet of fallen leaves on the forest floor is in a constant state of decomposition and replenishment, supplying a steady source of food for soil microbes and trees. On virgin prairies, the dead grasses from last year's growth perform a similar function, providing a mat of decaying organic material to sustain current growth.

Clearing the forest or plowing the sod is a necessary step for cultivation of crops, but the conversion has its cost. Cultivation enables crop plants to secure nutrients at an accelerated rate and thus also accelerates the depletion of organic matter in the soil. In converting a piece of land from its natural state to one of cultivation, the farmer or gardener is putting into high gear the process by which decaying organic content is consumed. To keep this natural engine running at high speed, the fuel must be replenished — more organic matter must be supplied to maintain a high state of fertility.

Profitable production of crops can only be temporary unless some provision is made for replenishing organic matter. History provides numerous examples. Colonists in Connecticut as early as 1750 were abandoning fields that had fallen in production after centuries of leaf mold had been consumed in just a few short years of corn production. The organic content of virgin soil represents centuries of deposits. The colonists withdrew those deposits quickly through a hastened use of the organic matter. Unless gardeners make new deposits along with the withdrawals, the account is soon so low that dividends — in the form of lush crops — disappear.

Food for the Long Haul

Gardeners who truly wish to improve their soil must shun the "fast-food" mentality of throwing on enough fertilizer to get by for this year and instead concentrate on the long-term goal of building a garden soil that will be productive for years. Just as the soil was not created overnight, improving poor or depleted soil with additions of organic materials can take several years. While "fresh" and partially decomposed plant material provides some benefit, even more benefits come after the material has been in the soil long enough to reach the advanced stage of decomposition, humus.

A soil with abundant humus saves you money because it needs less applied fertilizer and less applied lime (to raise soil pH of acidic soils) or sulfur (to lower soil pH of alkaline soils). Plants growing in such soils can tolerate higher or lower pH than can plants growing in depleted soils; they can also better withstand the stresses of drought, heat, and cold. But in addition, these soils do a better job of holding on to any fertilizers or lime that you apply; these won't wash away in the first rainstorm.

Different soil types and different climates affect the amounts of organic matter present in both virgin and cultivated soils. People who garden on sandy soils or live in warm climates will have a harder time maintaining high levels of soil organic matter. Warm temperatures increase the rate of decomposition, so an inch of organic matter spread over soil in a cool climate will last at least twice as long as the same material spread in a warm, humid climate. In natural desert environments, the decomposition rate is much slower, but supplying enough water to grow vegetables speeds up the process considerably.

It's hard to determine precisely how much organic material is present; soil tests usually estimate it based on the carbon content of a soil. This, plus the effects of different soil types and climates,

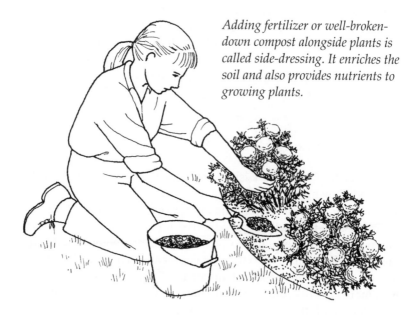

Adding fertilizer or well-broken-down compost alongside plants is called side-dressing. It enriches the soil and also provides nutrients to growing plants.

makes it unrealistic to talk specific numbers. While bringing organic matter levels up to 2 percent might be an accomplishment for sandy soils in the southeastern United States, the same amount would be quite inadequate for clay soils in the cool Northeast or Northwest.

Even when organic material is added on a regular basis, the levels in cultivated soils will be lower than those of virgin soils in the same area because cultivation increases the rate of decomposition. As a result, it's essential to replenish organic matter on a regular basis, to make this part of your gardening routine.

Organic matter can be added simply by spreading it on the soil surface. If the material is well decomposed such as fine finished compost, this is called top-dressing (or side-dressing, if it is spread alongside rows of vegetables or around individual plants). Well-decomposed top-dressings go to work almost immediately; earthworms will help mix them into the soil. If material spread on the soil isn't as well decomposed, it's called mulch. When organic material such as chopped leaves or manure is turned into the soil, it's called a soil amendment. Green manures are plants grown for the purpose of being turned under the soil to build up organic matter; growing a green manure is the cheapest and easiest way to improve soil over a large area. Mix and match these different methods to create a program that suits your available time, energy, and budget.

If your soil is chock-full of earthworms, consists mostly of loose crumbs the size of peas or marbles, and is easy to work, you know it's got about the right amount of organic matter. Congratulate yourself and keep up the good work. If it doesn't have these characteristics, step up your efforts using the techniques explained here.

Soil Amendments Close to Home

Any substance that will rot is a suitable addition to garden soil. If it's dug in for the purpose of improving the soil's structure or organic matter content, it's called an amendment or soil conditioner. (If it's added because it's a particularly good source of one or more plant nutrients, call it a fertilizer instead.) Excellent low-cost soil amendments can be found in your yard or nearby in your neighborhood. Spread the fresh materials over the soil and mix them in with a garden fork or rotary tiller. The same materials also make good additions to a compost pile.

Fallen leaves are one of the best amendments. Deep-rooted trees draw up minerals from deep in the ground that aren't available to relatively shallow-rooted vegetables. If turned under soon after they fall, leaves contribute nitrogen as well as other nutrients; if allowed to form leaf mold before spreading on soil, they'll be lower in nitrogen but still an excellent soil conditioner. Chop leaves with a lawn mower or shredder so they'll break down and release nutrients quickly. If you're out for a drive in the fall and see someone bagging up leaves, stop and ask if you might have them. At worst you might be taken for an oddball. At best, you'll secure some free organic matter for your garden.

The garden is an excellent source of organic materials. Weeds are a nuisance, but if allowed to grow to just before they set seed they can be tilled under to provide a good source of organic material. Many weeds are deeper rooted than vegetable plants and thus have greater access to subsoil minerals. When weeds are tilled under and decompose, these accumulated minerals become available to the crops that follow.

The inedible portions of vegetable plants are another storehouse of organic material. The vines of cucumber and tomato plants, the stalks and leaves of broccoli and cauliflower, and carrot tops — all these and more should be returned to the soil. Add them to the compost pile, or dig a trench in a portion of the garden that won't be used for a few months to compost them in place. If your garden is heavily mulched, you can tuck vegetable wastes underneath the mulch.

Your kitchen provides other organic matter. Coffee grounds and tea leaves can be spread right on the soil. Vegetable and fruit peelings are more unattractive and so are best added under mulch or placed in the compost pile.

Folks in the country can sometimes obtain manure for free by offering to clean out a neighbor's horse stable, chicken coop, or pigpen. Most farmers now charge a good amount for manure, but they'll deliver it by the truckload. If you don't like cleaning out stables, that's a worthwhile investment. Fresh manure can burn growing plants, so dig it in at the end of the growing season. Let it age or compost it before spreading it around growing plants.

If you live near the coast, seaweed is free for the gathering. It's a valuable source of trace minerals as well as readily available organic matter. You may prefer to use it as mulch rather than digging it into the soil.

Investigate manufacturing activities in your area. If they're not already selling their processing waste to a municipal composting facility, they may be willing to give some to you. Cannery wastes are a good source of organic matter. Pea and bean pods, potato skins, corncobs, peanut shells, and the like will all boost soil fertility. Winery pomace, apple pomace from cider pressing, and brewer's waste from beermaking are also good, but they're harder to handle because they're wet. The waste from the produce section of a local supermarket can be a fortune in free organic materials for you if you provide your own container and collect the refuse regularly.

Sawmills can supply you with sawdust; you may have to pay for it because it's often sold as animal bedding. Carpenters can also supply sawdust or wood shavings. Keep in mind that sawdust and wood shavings are deficient in nitrogen, so you need either to allow them to weather for a few months before using or to mix them with a good source of nitrogen such as manure or blood meal (or standard fertilizer). Sawdust is an easy mulch to apply evenly around plants and makes for an attractive garden.

Growing Your Own

One of the best ways of improving the soil's organic content with a minimum of time, effort, and expense is through the use of green-manure crops. These are plants grown for the express purpose of plowing under at an immature stage of growth to boost the organic content of the soil. If also grown to reduce erosion, such as planting in fall to keep garden soil covered through the winter, they may be called cover crops. Scores of different plants are used for this purpose; one is sure to be suited to your particular climate, type of soil, and planting schedule. Crops grown for organic matter can be divided into two categories: legumes and nonlegumes.

Legumes are plants that form symbiotic relationships with specialized soil bacteria that (unlike plants) are able to use nitrogen from the atmosphere, which is present in soil air. The bacteria are able to convert this otherwise inaccessible nitrogen into a form plants can use through a process known as nitrogen fixation. Where these attach themselves to the roots of legumes, little nodules form; you can see the tiny bumps on the roots if you pull up healthy legume plants.

To contribute significant amounts of nitrogen, most legumes need to be grown for at least a year (preferably two) and turned under the following growing season. (They can be mowed several times to produce hay for mulch or compost.) When mature legumes are plowed under the soil and decompose, the nitrogen is released for use by other plants and soil organisms. The nitrogen content of a mature stand of alfalfa or hairy vetch can be 3 to 4 percent (as much as some organic fertilizers, which may contain alfalfa meal as an ingredient). That's about three times the maximum content of a nonlegume such as oats or rye.

For the gardener, legumes are among the most valuable crops. In addition to contributing nitrogen to the soil, many such as peas, soybeans, and beans double as nutritious vegetables. Genetic engineers are also eager to exploit the alchemy that allows legumes to extract free fertilizer from the atmosphere. Work is being conducted to try to splice the genes responsible for nitrogen fixation onto plants such as corn and cereal grains that require high fertility. For a hungry world in which most nitrogen fertilizer is produced from natural gas, this is valuable research.

Nonlegumes often grow more quickly than legumes. They overtake and shade out many annual weeds, making them useful

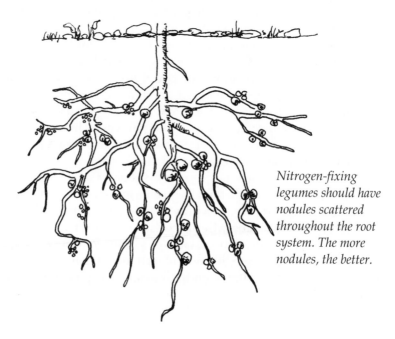

Nitrogen-fixing legumes should have nodules scattered throughout the root system. The more nodules, the better.

for weed control. With the exception of buckwheat and rapeseed (canola), these are all grasses. In addition to benefiting soil, these make high-quality feed for livestock. Green manures can be mixtures of plants; a majority of the hay crops grown in this country contain at least some percentage of legume forage.

Benefits of Green Manures

Green manures are economical. It takes half a pound or less of seeds to plant a garden of 100 square feet, which generally costs only a dollar or two. The more you buy, the cheaper the price per pound — so green manures become even more cost effective for large gardens and farms.

rye

Green manuring is easy. Seeds can be mail-ordered if you don't have a farm-supply store nearby. Sowing seeds takes very little time, and the immature crop can be tilled into the soil in short order. Small patches can be turned under using a garden fork. A couple of handfuls of seed produce as much organic matter as several bushels of well-rotted manure or compost, and they're much easier to spread. It's much easier to grow the organic matter in place exactly where you want it.

Legumes Need Inoculants

Legumes can't fix nitrogen unless the right bacteria are present. When you buy legume seeds, it's important also to purchase a supply of bacteria — called an inoculant. The combination inoculant sold for garden peas and beans won't work for most legumes grown as green manures. Vetches and broad beans require another type (species) of inoculant, soybeans another, and clovers still a different one. Ask your seed supplier for inoculant that matches your seed type. Follow label directions, which will specify whether the powder should be mixed with seeds or sprinkled into garden rows.

Green manures help to fight weeds in the garden. If seeded at a relatively thick rate, fast-growing types such as buckwheat will create a dense stand that will grow vigorously enough to deny light, moisture, and nutrients to weeds. Winter rye is particularly good; it produces chemicals toxic to many weed seedlings, so one crop can significantly cut down on weeds. Even tough perennials such as quack grass will succumb to successive plantings of green manures.

buckwheat

The organic matter from green manures — particularly legumes — is more rapidly available to plants than that from many other types of soil amendments. Turned under when stems and leaves are still tender and lush, the plants decompose rapidly. (The nitrogen added by legumes hastens this process even more.) Rapid breakdown of the organic matter means that succeeding crops and beneficial soil organisms will benefit quickly from the nutrients being liberated.

Fitting into the Garden

If you have space enough to expand your current garden, consider having two plots. While one is producing vegetables, the other can be used for growing soil-improving green manures. Alternating back and forth will build the soil in each bed and at the same time reduce pests and diseases (growing the same crop in the same soil year after year will encourage insect and disease populations to build up).

If you can't double your current space, plant green manures before or after other crops, or even between rows during the regular gardening season. An easy way to include green manures is to plant early produce in a different location within the garden each year. Peas, radishes, and lettuce can be planted as soon as the ground is workable and grow quickly, so they're all out of the ground relatively early in the gardening season. After they're harvested, turn under their residue, level a good seedbed with a rake, and plant a green-manure crop.

Types that grow well in cool temperatures can be planted toward the end of the growing season as garden beds are emptied; turn them under in spring a couple of weeks before planting other crops. Winter rye and hairy vetch are very hardy and will survive cold winters; you can plant these right around late crops such as broccoli. Oats are killed by heavy frost but are still useful because of their rapid growth rate. If sown in September, they'll produce enough growth that the frost-killed plants will do a good job of covering the ground over winter to reduce erosion.

Green manures can also be planted among certain vegetable plants. It takes a long time for cucumbers to grow from seed to sprawling vines. Take advantage of that period to sow a short-season green-manure crop that can be turned under before the vines run. Or, if your garden paths are wide enough for a lawn mower, sow white clover in them and keep it short by mowing periodically.

A rotary tiller is a tremendous help in turning under large stands of green manures.

Best Legumes for Green Manures

Alfalfa. The king of the soil-building crops, alfalfa is the best nitrogen fixer and will grow roots as deep as 20 feet if left in place for two years. It requires fertile, neutral to alkaline soil and won't tolerate wet growing conditions. Perennial; adapted to all regions. (¾ oz./100 sq. ft.; ½ lb./1000 sq. ft.)

Alsike clover. This biennial, best adapted to northern and central states, fares better on poorly drained or acidic soils than most clovers. Sow in spring and turn under in the fall, or sow in late summer and turn under the following spring. (¾ oz./100 sq. ft.; ½ lb./1000 sq. ft.)

alfalfa

Cowpea. An annual, this makes fast growth if sown after soil has warmed up; it's best for southern and central states. It tolerates drought, light shade, and poor soil (as long as it's well drained). Like other deep-rooted crops, it improves aeration and roots reach into subsoil for nutrients, making them available for succeeding crops. (4½–6 oz./100 sq. ft.; 3–4 lb./1000 sq. ft.)

Crimson clover. An annual, this needs neutral pH but tolerates poor soils. Where winters stay above 10°F, sow in fall and turn over in spring; in cold climates sow in spring and turn over in fall. (¾–1½ oz./100 sq. ft.; ½–1 lb./1000 sq. ft.)

alsike clover

Fava (broad) beans. This annual produces edible beans (any edible beans can double as green manure if turned under after harvest). Prefer cool weather; plant at the same time as garden peas and till under in summer or after harvest. (6 oz./100 sq. ft.; 4 lbs./1000 sq. ft.)

Field peas. Annual field peas tolerate a wide climatic range and many types of soils, including clay. They work best if combined

crimson clover

with oats or other grain; sow in spring and turn under before peas fill out pods or, where winters are mild, sow in fall and turn under in spring. (4½–6 oz./100 sq. ft.; ¾ lbs./1000 sq. ft.)

Lespedeza (bush clover). All lespedezas are good for restoring eroded soils or improving poor ones in warmer climates. They tolerate acidic soils. Though one type is perennial, all are treated as annuals by planting in spring and tilling under in late summer. (1½ oz./100 sq. ft.; 1 lb./1000 sq. ft.)

Red clover. This perennial prefers the cooler temperatures of northern states. Sow in early spring. This crop is exceptionally fast to decay and thus of more immediate benefit to following crops. (¾ oz./100 sq. ft.; ½ lb./1000 sq. ft.)

lespedeza

Soybeans. Annual soybeans need warm weather; sow after soil has warmed and turn under in summer or fall. If growing for organic material rather than the beans themselves, select a long-season variety, as these will make the largest plants. (6 oz./100 sq. ft.; 4 lbs./1000 sq. ft.)

Sweet clover. Annual varieties prefer cooler climates; biennial types are good for all except the Gulf states. These tall plants develop deep roots that can penetrate difficult soils and hardpan; they tolerate less fertility than alfalfa and produce almost as much nitrogen. Sow biennials in spring and turn under the following spring to secure maximum growth and nitrogen fixation. (¾ oz./100 sq. ft.; ½ lb./1000 sq. ft.)

red clover

Vetch. Most are biennials suited to all areas with reasonably good soil and adequate moisture. Hairy vetch is the only one hardy enough to withstand severe winters; it also tolerates poor and/or acidic soils and is the best choice for short-term nitrogen fixation. For good winter cover, mix hairy vetch with grains such as rye and oats. Sow in spring or fall and till under in fall or the following spring. (2½ oz./100 sq. ft.; 1½ lbs./1000 sq. ft.)

White clover. This perennial is shorter than most green manures and tolerates foot traffic, so it can be grown in garden paths or around other crops and seeded in lawns. Sow spring to summer and turn under in fall or spring. (¼–½ oz./100 sq. ft.; ¼ lb./1000 sq. ft.)

Planting Tips

Select a green manure suited to your climate, soil, and the season you intend to grow it. Buy fresh seeds, labeled for the year in which they're to be grown. The seeding rate varies depending on the species; check the descriptions (see boxes) to determine how much you need.

Only a few green manures such as buckwheat and crimson clover tolerate poor soil. To produce the lush, abundant growth needed to

---

Best Nonlegumes for Green Manures

Barley. Grow spring varieties of this annual in the North and winter varieties in hot climates with mild winters. It needs neutral to alkaline soil that isn't too sandy. (4 oz./100 sq. ft.; 2½ lbs./1000 sq. ft.)

Bromegrass. This perennial is among the hardiest and easiest to grow of the soil-building crops. Sow in early spring or late summer throughout the northern half of the U.S.; planted in the fall, it's a good winter cover. (1½ oz./100 sq. ft.; 1 lb./1000 sq. ft.)

Buckwheat. For rebuilding neglected, poor soil, this annual is unsurpassed; it tolerates acidic soils also. Sown after the ground is warm, buckwheat will be ready to plow down in six weeks. This vigorous top growth coupled with a massive root system can add as much as 40 tons of green matter per acre per summer. Buckwheat is good for smothering weeds and is able to use plant foods (especially phosphorus) locked up in the subsoil far better than most crops. (3–4½ oz./100 sq. ft.; 2–3 lbs./1000 sq. ft.)

bromegrass

Millet. This fast-growing annual is good for smothering weeds; it tolerates acidic soils and drought. It needs warm weather (not frost hardy), so plant in late spring or summer and till under in summer or fall. (1½ oz./100 sq. ft.; 1 lb./1000 sq. ft.)

buckwheat

supply abundant organic matter, most types require the same soil needed to grow good vegetables. Test the pH and nutrient levels of your soil to see if you need to add lime, sulfur, or fertilizer before you plant. Add these before you till, or rake into the top few inches of soil. Once your soil is in good shape from several years of improvements, you may not need to add anything to grow good green manures.

Green manure seedbeds don't need to be prepared as finely as for vegetable seeds. Since covering each seed is impractical, some

Oats. Seeds of this annual can germinate in cool soils, so it's among the earliest crops that can be sown by northern gardeners. If given even moisture, it grows quickly and can be turned under in six to eight weeks. Fall sowings will be winter killed in all but mild climates (but dead foliage still provides some protection against soil erosion). (¾ oz./100 sq. ft.; ½ lb./1000 sq. ft.)

Rapeseed (canola). This fast-growing biennial turnip relative produces edible greens. In northern and central states, plant spring types early in the season and turn under in summer before pods form. In the South, plant winter types in the fall for spring tilling. (7½ oz./100 sq. ft.; 5 lbs./1000 sq. ft.)

Rye. Rye comes in two forms; both are fairly adaptable. Annual rye (ryegrass) is fast growing and will die as the first freezes come in fall (but dead growth still provides some protection against soil erosion). If planted early in the season, turn it under before plants bloom or form seeds, as it can be weedy. Perennial winter rye is very hardy and can be sown well into fall; it will put up new growth in spring. Winter rye is great for weed control; it should be turned under three to four weeks before planting any other seeds. (annual: 1½–3 oz./100 sq. ft., 1–2 lbs./1000 sq. ft.; perennial: 4½–6 oz./100 sq. ft., 3–4 lbs./1000 sq. ft.)

rye

Weeds. Where they don't compete directly with other crops, weeds can add valuable organic matter. Most are nonlegumes. Turn under before seeds form.

roughness of the soil surface is desirable so that raking or tamping will produce adequate coverage.

For small areas, broadcast seeds by hand. First practice the motion of sowing with fertilizer or some substance like flour. Try to swing your arm and wrist to get good coverage and spread.

For larger areas, you may want to buy a spinner spreader that can be slung over your shoulder. Operated by a hand crank, this spreader holds seeds or fertilizer in a bag. The material passes through an adjustable gate onto a spinner, which flings the material evenly over the surface of the garden.

To provide even coverage and thus a good stand of plants, divide the total quantity of seeds into halves. Sow half the seeds in one direction, then sow the other across the first seeds perpendicular to the first direction.

If seeds are very large, such as peas and beans, rake them into the soil for good coverage. For fine seeds such as clover and alfalfa, tamping with the teeth of a rake or with a lawn roller (nearly empty) will suffice.

Unless the soil is quite damp, hose down the seeded area with a fine mist. Most green manures should sprout in less than two weeks if soil conditions are good. The crop should need no care. If a lot of weeds sprout among slower-growing perennial legumes, mow or cut your planting once it reaches a few inches high; the perennials will bounce back quickly and overtake the annual weeds.

Returning Green Manures to the Soil

Turn under the crop when it is actively growing, tender, and succulent. For fast-growing types such as buckwheat, this is in about six weeks or as soon as the first flowers appear. Never allow the green manure crop to go to seed. A weed has been defined as a plant out of place, and it holds true for buckwheat or alfalfa in addition to crabgrass.

If plants start to bloom and you don't have time (or the soil is too wet) to turn under the crop, mow or cut the plants to prevent seed formation and buy a little time. Cutting or mowing the crop before turning it under often makes the job easier. Try a string trimmer or grass clippers for small plots.

Perennial legumes such as alfalfa will supply the most nitrogen if left to grow more than one year. If you're growing legumes for a year or two, mow periodically to keep stems from becoming too tough to cut or turn under easily. Decomposition will be greatly slowed if plants are allowed to become woody, and fresh plant material decays much more quickly than dry material. Trimmings can be used as mulch for vegetables elsewhere in the garden, spreading the wealth. The extensive root systems of green-manure plants will soon produce new, tender top growth.

Mulches

Another excellent method of improving the soil with minimum effort is to mulch the garden. Growing or gathering your own materials takes a little more effort but costs you nothing. A mulch is a layer of material spread on the surface of the soil to retain moisture and retard weed growth. Only organic mulching materials contribute humus to the soil; inorganic materials such as landscape fabric, gravel, and plastic mulch don't. Black plastic stops weeds, warms soil, and conserves moisture, but it adds nothing to the soil.

Organic mulches don't increase soil fertility or humus as rapidly as digging in amendments or green manures, but they're nonetheless good soil builders. Mulching the garden imitates nature. The layer of decaying leaves on the surface of the forest soil is a mulch, as is the layer of matted, decaying grass in grasslands.

Mulches greatly cut down on the need to weed and water the garden. They also help protect the soil from temperature extremes,

Leaf Mold

It's easy to make your own leaf mold, which is an excellent mulch and will be welcomed by the earthworms in your garden. Simply pile chopped leaves in a corner of the yard, and in a year you'll have leaf mold. Chop leaves with the lawn mower or a leaf shredder. If your yard is windy or you like things to look tidy, fence in the pile with chicken wire or a cage of cement-reinforcing wire.

If you don't chop the leaves first, they'll take a couple of years to break down. Also, you'll need to start a new pile each year, or the material buried at the bottom will take even longer to decompose.

keeping the ground warmer in the winter and cooler in the summer. They can prevent sprawling plants such as cucumbers and tomatoes from coming in contact with bare ground, precluding many rot and fungus problems and keeping mud off salad greens and other leafy vegetables. Soil erosion from wind or water is minimized in a mulched garden.

You can purchase bagged mulches such as shredded bark, cocoa hulls, and bark nuggets at any garden center; if you use a lot, order by the truckload from a local nursery or landscaper. But it's much cheaper to gather your own materials close to home.

Leaves should be chopped before use (with a shredder or lawn mower) to keep them from forming a water-repelling mat; grass clippings should be dried first for the same reason. Fresh sawdust, wood chips, and shavings should be allowed to weather outdoors (or spread a good nitrogen source such as manure, blood meal, or standard fertilizer) before using, as they require a lot of nitrogen to decompose and you need to provide extra to keep your plants well supplied.

Old newspapers, laid a few sheets thick between vegetable rows, are a fine mulch and will decompose rapidly. Worms love them. Avoid using the four-color advertising sections. If the sight offends you, cover newspapers with a layer of straw, hay, or chopped leaves.

In vegetable gardens, mulches can double as soil amendments if turned into the soil at the end of the gardening season. Or leave over winter to protect soil and turn under the following spring. Use a soft material that will break down quickly. Once turned under, microbes can work more rapidly, making the organic material available as plant food.

In perennial flower gardens and around permanent crops such as berry bushes and asparagus, leave the mulch in place year after year, replenishing the protective covering as it starts to decay. After a few years, this practice will yield notice-

Free or Inexpensive Mulch Materials

Chopped leaves

Compost

Grass clippings

Ground corncobs

Leaf mold

Newspaper

Pine needles

Sawdust

Seaweed

Spoiled hay

Straw

Wood chips or shavings

able benefits. Soft materials such as compost will benefit soils more quickly, but they also have to be replenished more often. Harder materials such as wood chips last longer.

Mulched soils are very slow to warm up in the spring, but the mulch can be pulled to the side once the danger of hard frost has passed, then put back in place once plants begin to grow. Another disadvantage in humid climates is that mulches can attract slugs and snails or keep soil too moist, encouraging some diseases. Keep mulches a few inches from the crowns of perennial plants to discourage diseases.

Composting

Compost is the best form of organic matter for improving soil. Like other sources of organic matter, it supplies small amounts of all the major nutrients in a balanced, slow-release form and helps soils hold on to nutrients and moisture long enough for plants to use them. It supplies all the micronutrients, too. But in addition, it inoculates your soil with beneficial microorganisms; these are important for making nutrients available to plants and can suppress disease organisms in soil. Compost also provides the best food for these beneficial organisms.

While it can be bought by the bag at any garden center, or by the truckload, compost is so easy that you might as well make it yourself. Early in the spring, start collecting whatever organic materials you can get your hands on: weeds, leaves (chop with lawn mower), garden trimmings, kitchen scraps, manure, and anything else mentioned in the section on soil amendments. If the material is large, bulky, or very stiff, cut or chop it to help it break down; later, that will save you from pulling large pieces out of otherwise-finished compost. If you pile these materials in a corner of the yard, you'll eventually get good compost even if you rarely (or never) turn them.

Avoid diseased plants and weeds with seeds (or weeds that can sprout from bits of roots), as your pile may not get hot enough to kill these. Fats, oils, and grease keep anything they coat from breaking down, plus large amounts attract animals; meat scraps also attract animals. Used kitty litter and pet feces may carry parasites and diseases that infect humans. Finally, don't put anything in compost that won't break down, such as glass, pressure-treated wood, charcoal briquettes, or plastic.

Volumes have been written on the subject of composting, and it can be made as simple or as complicated as you wish. Some gardeners build fairly elaborate permanent structures for their compost piles, while others simply use the open ground. Some insist on a precise recipe of materials, while most will toss in any material at hand. Whatever the method, composting affords the home gardener a good way to convert organic materials that might otherwise be wasted into valuable plant food and soil conditioner.

Pick a good spot for your compost pile. It should be handy to the kitchen and garden, and preferably not the center of attraction or right below a window. A bin is optional. With or without a bin, the pile should be about 4 feet square and 4 feet high. One of the easiest bins is a cylinder made from a 10-foot strip of wire mesh that is 4 feet wide; to turn the pile, just pull off the bin, set it alongside, and toss the old pile into the now-empty bin. Or build a framework that can be taken apart, and pull off the front panel to turn the contents.

For the fastest compost, watch your carbon-nitrogen balance, turn piles with a garden fork every week to 10 days, and keep the contents evenly moist but not soggy. By the end of a month to six weeks, the heap should be greatly reduced in volume and its ingredients largely decomposed. Covering piles keeps rain from soaking them in wet climates and helps keep them moist in dry climates. If

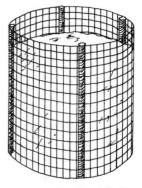

Two easy-to-construct composters are made from stacked cement blocks, below, and wire secured to stakes driven into the ground, at right.

A Recipe for Compost

To get the best humus-rich compost, mix materials to get a good balance of carbon and nitrogen. A mixture with too much carbon (such as piles of leaves) won't heat up and will take longer to decompose, while a mixture with too much nitrogen will get smelly and slimy.

Think of high-carbon materials as "browns" and high-nitrogen materials as "greens" (this works for most things, though manure is high in nitrogen but not green). Use 3 parts loose, bulky, high-carbon material for each part fresh high-nitrogen material. (If your nitrogen source is dry and relatively fine, such as dehydrated manure or blood meal, mix it with 4 to 6 parts bulky carbon material.) You can add another part of materials such as sod that contain a balanced amount of carbon and nitrogen. (Sod is also an excellent source of decomposer organisms.) If you have a shredder, you can measure and mix at the same time as you run material through the machine. Another easy way to measure is to layer the materials:

Top 6 inches of browns with 2 inches of greens and up to 2 inches of balanced material; repeat until the pile is 4 feet high or the compost bin is full. Moisten each layer as you go and turn after a week or two to mix all the materials thoroughly.

GREENS	BROWNS	BALANCED
Fresh grass clippings	Straw	Manure with bedding
Fresh or dehydrated manure	Hay	Well-aged manure
Human hair	Leaves	Pea and bean pods
Alfalfa hay or meal	Sawdust	Fruit peels, cores
Crushed eggshells	Wood shavings	Vegetable peelings
Cabbage, broccoli leaves	Shredded newspaper	Sod and soil
Sour milk	Dry grass clippings	Fresh weeds
Feathers, wool	Cornstalks	Rotted wood
Kelp meal, seaweed	Peanut shells	Soft, green garden trimmings
Apple or winery pomace		

you don't want to turn piles, aerate periodically by poking a long pole or steel bar down to the bottom and wiggling it around to make holes to let in air. Compost is finished when most ingredients have become a crumbly or fluffy, dark brown, soil-like material with the sweet, woodsy smell of rich soil.

Using Compost

If your supply of compost is ample, or if you want to use rough compost that hasn't finished breaking down, spread up to 3 inches on the surface of the soil and work into the top few inches. Or leave it on top of the soil over the winter to reduce erosion; even rough compost will be finished by spring.

If your supply is limited, use it only in the immediate vicinity of the plants as a top-dressing or mulch. Since finished compost

---

Grass Clippings

Lawn clippings are a good source of organic material. They're equally beneficial left in place on the lawn (especially when finely chopped by today's mulching lawnmowers). But if your interest in garden produce is greater than in greensward, the mowings can be bagged or raked up and applied to the garden soil.

Don't leave fresh clippings in a pile, or they'll quickly turn to a smelly brown mess. Don't pile them too thickly, either, if you're using them as mulch or in the compost pile, or the same thing will happen. Here are several ways to use them:

- Add them to the compost pile. Fresh clippings will give you the nitrogen you need to make the pile "cook." Mix them well with other materials such as hay, chopped leaves, and weeds. If herbicides are used on the lawn, compost clippings before using them in the garden (ideally, avoid herbicides if you're using the clippings).
- Spread a layer on the soil, then till them in. They're an excellent soil amendment.
- Spread them in thin layers when green, or let them dry before spreading them in the garden, and they'll provide one of the best mulches you can find.

---

releases nutrients slowly, it can't burn growing plants. Add a handful to each planting hole for transplants; add a shovelful for larger plants and heavy feeders such as tomatoes, melons, and squash. Sprinkle compost over rows after planting seeds. Feed established plants by laying down a band of compost along the row as a side-dressing. To use as a spring or fall top-dressing for lawns, screen compost through ½-inch mesh to remove large pieces.

Fertilizer Possibilities

While low-cost soil amendments can form the core of your soil improvement program, sometimes a cash outlay is needed to make your garden fully productive. To get an accurate picture of what your soil needs, get a soil test. These are performed by private laboratories (more expensive but more detailed) or by your local Cooperative Extension Service. Once plants are growing well, a test every four years or so will keep you on track.

Wood Ashes

Don't overlook the final, valuable product from your fireplace or woodstove. Ashes from wood fires are a valuable source of potash (the common form of potassium). If you bought them as a commercial fertilizer, bagged and ready for use, the bag would be marked 0–2–7, indicating 2 percent phosphoric acid and 7 percent potash (actual nutrient content varies). To correct a potassium deficiency, use about 3 pounds of wood ashes for 100 square feet and work them into the top few inches of soil.

Remember that ashes are very alkaline, so use them only on acidic soils (pH 6.5 or lower) and don't add lime the same year. Don't spread them on soils for acid-loving plants such as potatoes and blueberries. To raise the pH from 5.5 to 6.5 on loamy soil, you need about 8 pounds of wood ashes; use 3–4 pounds for light, sandy soils and 9–10 pounds for heavy clay.

To get the most from your ashes, store them where the rain won't leach away the potash. A metal trash can is good (don't use flammable containers; house and barn fires have started from "dead" ashes).

Test results will indicate the acidity or alkalinity of the soil (the pH), and what you need to add to bring soil to the ideal pH range of 6.3 to 6.8 (slightly acidic). When soil is too acidic or alkaline, existing nutrients are locked up out of plants' reach. The ideal, slightly acidic pH promotes the maximum nutrient availability and is best for soil organisms (including those that fix nitrogen) and most plants. Add lime (for acidic soils) or sulfur (for alkaline soils) at rates recommended by the soil test, or your plants won't be able to make full use of any fertilizer you add. Use ordinary (calcitic) lime unless your soil test shows a magnesium deficiency; use dolomite to supply magnesium as well as calcium. You can substitute wood ashes for limestone (see box on page 27 for amounts); if so, you won't need to add any potassium that year, so avoid standard fertilizers such as 5-10-10.

Soil tests will also report the levels of major nutrients. If any are deficient, you can bring your soil into balance with organic fertilizers; tests will usually include recommendations for how much to add. You can buy these premixed in bags from any garden supplier.

PRIMARY PLANT FOOD ELEMENTS

ELEMENT	SYMBOL	FUNCTION IN PLANT
Nitrogen	N	Gives dark green color to plant; increases growth of leaves and stems; influences crispness and quality of leaf crops; stimulates rapid early growth.
Phosphorus	P	Stimulates root formation and growth; gives seedlings rapid and vigorous start; essential for formation of flowers and seeds.
Potash	K	Increases overall vigor, disease resistance, and cold hardiness; stimulates production of strong, stiff stalks; improves quality of crop yield; promotes production of sugar, starches, oils; increases plumpness of grains and seeds.

While organic fertilizers have lower N-P-K formulas than standard synthetic fertilizers, they supply essential nutrients such as calcium, magnesium, sulfur, iron, and micronutrients missing in standard synthetic formulas. Most organic fertilizers also supply some organic matter, and they release their nutrients slowly over a longer period so there's no danger of burning plants or soil organisms.

The most concentrated forms of natural fertilizers usually have to be purchased (see chart below). But many materials close at hand also supply plant nutrients, and a number of these are worth adding to soil just for the organic matter they supply. If you develop a scavenger mentality so you're always on the lookout for materials to recycle through your soil, sources will suddenly seem to appear.

Manures vary greatly in their nutrient content, but all are rich sources of calcium, micronutrients, and organic matter. Fresh manure is one of the best additions to a compost pile. You can also spread 2 to 4 inches of manure on the soil at the end of the fall growing season (keep fresh manure away from actively growing plants). If you don't have access to stables or farms, you can buy dehydrated

DEFICIENCY SYMPTOMS	EXCESS SYMPTOMS
Plants fade to light green or yellow; new leaves and shoots smaller than normal; slow or stunted growth.	Dark green; excessive growth; delayed maturity; fewer (or lower quality) flowers, fruits, or tubers; lowered resistance to stress and diseases.
Red or purple areas on older leaves; fewer flowers or fruits than normal; plants grow too slowly.	Micronutrient deficiencies, which are hard to identify but usually involve yellowing and/or odd growth (phosphorus can tie up other essential elements).
Patchy yellow or dead spots on older leaves, spreading to new leaves; reduced vigor; increased susceptibility to diseases; weak or unusually short stems that may flop over; small fruits with thin skins.	Coarse, poor colored fruit; reduced absorption of magnesium and calcium.

manures in bags at garden centers; apply about 15 pounds per 100 square feet. For stronger types of manures (sheep, poultry, bat guano), use half as much. Composted manures can be applied at slightly higher rates.

If you're trying to correct a specific deficiency, choose materials higher in that nutrient. Good nitrogen sources include dried blood, fish emulsion, cottonseed meal (acidifies soil), and manure. The best phosphorus sources include bonemeal, fish meal or fish emulsion,

APPROXIMATE COMPOSITION OF NATURAL FERTILIZER MATERIALS

MATERIAL	NITROGEN (N)	PHOSPHORIC ACID (P)	POTASH (K)
PLANT AND ANIMAL CONCENTRATES			
Bonemeal, steamed	2.0	11.0	0
Corn gluten	9.0	0	0
Cottonseed meal	6.0	2.5	1.0
Dried blood meal	12.0	1.5	0.5
Eggshells	1.2	0.3	1.2
Feathers	15.0	0	0
Fish meal	5–9	3–7	2–5
Fish emulsion	4–9	1–7	1–2
Hair	12–16	0	0
Hoof & horn meal	10–15	2.0	0
Kelp meal (seaweed)	1.5	0.5	2.5
Sewage sludge (composted)	6.0	2.0	0
Soybean meal	7.0	0.5	2.3
Tankage	7.0	6.0	0–1.5
Wood ashes	0	2.0	7.0

and rock phosphate. Hard rock phosphate contains the most phosphorus but releases it very slowly; more expensive colloidal rock phosphate contains less but releases it faster. For potassium (potash) deficiencies, try greensand, langbeinite (good for alkaline soil, sold as Sul-Po-Mag or K-Mag), granite dust, or wood ashes (only good for acidic soil). Bonemeal, wood ashes, and rock phosphate are also excellent sources of calcium and reduce the need for limestone on acidic soils.

Rock Powders

Greensand (glauconite)	0	1.4	4.0–9.5
Langbeinite	0	0	22
Rock phosphate	0	18–30	4.5

Manures

Bat guano	8.0	4.5	2.0
Cow manure, dried	2.0	2.3	2.4
Cow manure, fresh	0.5	0.2	0.5
Cricket castings	4.0	3.0	2.0
Horse manure, fresh	0.6	0.3	0.5
Pig manure, fresh	0.6	0.5	0.4
Poultry manure, dried	4.5	2.8	1.5
Poultry manure, fresh	1.1	0.9	0.5
Sheep manure, fresh	0.9	0.5	0.8

Bulky Organic Materials

Alfalfa meal	3.0	1.0	2.0
Coffee grounds	2.0	0.2	0.3
Compost, commercial	1.0	1.0	1.0
Grass hay	1.2	0.4	1.5
Peanut shells	1.5	0.2	0.5
Sawdust	0.2	0	0.2
Straw	0.6	0.2	1.1

Other Storey Titles You Will Enjoy

The Complete Compost Gardening Guide,
by Barbara Pleasant & Deborah L. Martin.
Everything a gardener needs to know to produce the best
compost, nourishment for abundant, flavorful vegetables.
320 pages. Paper. ISBN 978-1-58017-702-3.

The Gardener's A–Z Guide to Growing Organic Food,
by Tanya L. K. Denckla.
An invaluable resource for growing, harvesting, and storing
765 varieties of vegetables, fruits, herbs, and nuts.
496 pages. Paper. ISBN 978-1-58017-370-4.

Let It Rot!, by Stu Campbell.
Stop bagging leaves, grass, and kitchen scraps, and turn
household waste into the gardener's gold: compost.
160 pages. Paper. ISBN 978-1-58017-023-9.

Starter Vegetable Gardens, by Barbara Pleasant.
A great resource for beginning vegetable gardeners:
24 no-fail plans for small organic gardens.
192 pages. Paper. ISBN 978-1-60432-529-2.

The Vegetable Gardener's Bible,
2nd edition, by Edward C. Smith.
The 10th Anniversary Edition of the vegetable gardening classic,
with expanded coverage of additional vegetables, fruits, and herbs.
352 pages. Paper. ISBN 978-1-60342-475-2.

The Veggie Gardener's Answer Book, by Barbara W. Ellis.
Insider's tips and tricks, practical advice, and
organic wisdom for vegetable growers everywhere.
432 pages. Paper. ISBN 978-1-60342-024-2.

These and other books from Storey Publishing are available
wherever quality books are sold or by calling 1-800-441-5700.
Visit us at *www.storey.com*.